THE
TWO CULTURES
AND
THE SCIENTIFIC
REVOLUTION

By C. P. Snow

THE REDE LECTURE • 1959

Martino Publishing
Mansfield Centre, CT
2013

Martino Publishing
P.O. Box 373,
Mansfield Centre, CT 06250 USA

ISBN   978-1-61427-547-3

© *2013 Martino Publishing*

Cover design by T. Matarazzo

Printed in the United States of America On 100% Acid-Free Paper

THE
TWO CULTURES
AND
THE SCIENTIFIC
REVOLUTION

By C. P. Snow

THE REDE LECTURE • 1959

CAMBRIDGE UNIVERSITY PRESS

NEW YORK

PUBLISHED BY

THE SYNDICS OF THE CAMBRIDGE UNIVERSITY PRESS

Bentley House, 200 Euston Road, London, N.W. 1
American Branch: 32 East 57th Street, New York 22, N.Y.

©

C. P. SNOW

1959

FIRST AMERICAN EDITION 1959

CONTENTS

I

THE TWO CULTURES

It is about three years since I made a sketch in print of a problem which had been on my mind for some time.[1] It was a problem I could not avoid just because of the circumstances of my life. The only credentials I had to ruminate on the subject at all came through those circumstances, through nothing more than a set of chances. Anyone with similar experience would have seen much the same things and I think made very much the same comments about them. It just happened to be an unusual experience. By training I was a scientist: by vocation I was a writer. That was all. It was a piece of luck, if you like, that arose through coming from a poor home.

But my personal history isn't the point now. All that I need say is that I came to Cambridge and did a bit of research here at a time of major scientific activity. I was privileged to have a ringside view of one of the most wonderful creative periods in all physics. And it happened through the flukes of war—including meeting W. L. Bragg in the buffet on Kettering station on a very cold morning in 1939, which had a determining influence on my practical life—that I was able, and indeed morally forced, to keep that ringside view ever since. So for thirty years I have had to be in touch with scientists not only out of curiosity, but as part of a working existence. During the same thirty years I was trying to shape the books I wanted to write, which in due course took me among writers.

There have been plenty of days when I have spent the working hours with scientists and then gone off at night with some literary colleagues. I mean that literally. I have had, of course, intimate friends among both scientists and writers. It was through living among these groups and much more, I think, through moving regularly from one to the other and back again that I got occupied with the problem of what, long before I put it on paper, I christened to myself as the 'two cultures'. For constantly I felt I was moving among two groups—comparable in intelligence, identical in race, not grossly different in social origin, earning about the same incomes, who had almost ceased to communicate at all, who in intellectual, moral and psychological climate had so little in common that instead of going from Burlington House or South

Kensington to Chelsea, one might have crossed an ocean.

In fact, one had travelled much further than across an ocean—because after a few thousand Atlantic miles, one found Greenwich Village talking precisely the same language as Chelsea, and both having about as much communication with M.I.T. as though the scientists spoke nothing but Tibetan. For this is not just our problem; owing to some of our educational and social idiosyncrasies, it is slightly exaggerated here, owing to another English social peculiarity it is slightly minimised; by and large this is a problem of the entire West.

By this I intend something serious. I am not thinking of the pleasant story of how one of the more convivial Oxford greats dons—I have heard the story attributed to A. L. Smith—came over to Cambridge to dine. The date is perhaps the 1890's. I think it must have been at St John's, or possibly Trinity. Anyway, Smith was sitting at the right hand of the President—or Vice-Master—and he was a man who liked to include all round him in the conversation, although he was not immediately encouraged by the expressions of his neighbours. He addressed some cheerful Oxonian chit-chat at the one opposite to him, and got a grunt. He then tried the man on his own right hand and got another grunt. Then, rather to his surprise, one looked at the other and said, 'Do you know what he's talking about?' 'I haven't the least idea.' At

3

this, even Smith was getting out of his depth. But the President, acting as a social emollient, put him at his ease, by saying, 'Oh, those are mathematicians! We never talk to *them*'.

No, I intend something serious. I believe the intellectual life of the whole of western society is increasingly being split into two polar groups. When I say the intellectual life, I mean to include also a large part of our practical life, because I should be the last person to suggest the two can at the deepest level be distinguished. I shall come back to the practical life a little later. Two polar groups: at one pole we have the literary intellectuals, who incidentally while no one was looking took to referring to themselves as 'intellectuals' as though there were no others. I remember G. H. Hardy once remarking to me in mild puzzlement, some time in the 1930's: 'Have you noticed how the word "intellectual" is used nowadays? There seems to be a new definition which certainly doesn't include Rutherford or Eddington or Dirac or Adrian or me. It does seem rather odd, don't y' know.' [2]

Literary intellectuals at one pole—at the other scientists, and as the most representative, the physical scientists. Between the two a gulf of mutual incomprehension—sometimes (particularly among the young) hostility and dislike, but most of all lack of understanding. They have a curious distorted image of each other. Their attitudes are so

4

different that, even on the level of emotion, they can't find much common ground. Non-scientists tend to think of scientists as brash and boastful. They hear Mr T. S. Eliot, who just for these illustrations we can take as an archetypal figure, saying about his attempts to revive verse-drama, that we can hope for very little, but that he would feel content if he and his co-workers could prepare the ground for a new Kyd or a new Greene. That is the tone, restricted and constrained, with which literary intellectuals are at home: it is the subdued voice of their culture. Then they hear a much louder voice, that of another archetypal figure, Rutherford, trumpeting: 'This is the heroic age of science! This is the Elizabethan age!' Many of us heard that, and a good many other statements beside which that was mild; and we weren't left in any doubt whom Rutherford was casting for the role of Shakespeare. What is hard for the literary intellectuals to understand, imaginatively or intellectually, is that he was absolutely right.

And compare 'this is the way the world ends, not with a bang but a whimper'—incidentally, one of the least likely scientific prophecies ever made—compare that with Rutherford's famous repartee, 'Lucky fellow, Rutherford, always on the crest of the wave.' 'Well, I made the wave, didn't I?'

The non-scientists have a rooted impression that the scientists are shallowly optimistic, unaware of man's condition. On the other hand, the scientists believe that the literary intellectuals are totally lacking in foresight, peculiarly unconcerned with their brother men, in a deep sense anti-intellectual, anxious to restrict both art and thought to the existential moment. And so on. Anyone with a mild talent for invective could produce plenty of this kind of subterranean back-chat. On each side there is some of it which is not entirely baseless. It is all destructive. Much of it rests on misinterpretations which are dangerous. I should like to deal with two of the most profound of these now, one on each side.

First, about the scientists' optimism. This is an accusation which has been made so often that it has become a platitude. It has been made by some of the acutest non-scientific minds of the day. But it depends upon a confusion between the individual experience and the social experience, between the individual condition of man and his social condition. Most of the scientists I have known well have felt—just as deeply as the non-scientists I have known well—that the individual condition of each of us is tragic. Each of us is alone: sometimes we escape from solitariness, through love or affection or perhaps creative moments, but those triumphs of life are pools of light we make for ourselves while the edge of the road is black: each of us dies alone. Some scientists I have known have had faith in revealed religion. Perhaps with them the sense of the tragic condition is not so strong. I don't know. With most people of deep feeling, however high-spirited and happy they are, sometimes most with those who are happiest and most high-spirited, it seems to be right in the fibres, part of the weight of life. That is as true of the scientists I have known best as of anyone at all.

But nearly all of them—and this is where the colour of hope genuinely comes in—would see no reason why, just because the individual condition is tragic, so must the social condition be. Each of us is solitary: each of us dies alone: all right, that's a fate against which we can't struggle—but there is plenty in our condition which is not fate, and against which we are less than human unless we do struggle.

Most of our fellow human beings, for instance, are underfed and die before their time. In the crudest terms, *that* is the social condition. There is a moral trap which comes through the insight into man's loneliness: it tempts one to sit back, complacent in one's unique tragedy, and let the others go without a meal.

As a group, the scientists fall into that trap less than others. They are inclined to be impatient to see if something can be done: and inclined to think that it can be done, until it's proved otherwise. That is their real optimism, and it's an optimism that the rest of us badly need.

In reverse, the same spirit, tough and good and determined to fight it out at the side of their brother men, has made scientists regard the other culture's social attitudes as contemptible. That is too facile: some of them are, but they are a temporary phase and not to be taken as representative.

I remember being cross-examined by a scientist of distinction. 'Why do most writers take on social opinions which would have been thought distinctly uncivilised and démodé at the time of the Plantagenets? Wasn't that true of most of the famous twentieth-century writers? Yeats, Pound, Wyndham Lewis, nine out of ten of those who have dominated literary sensibility in our time—weren't they not only politically silly, but politically wicked? Didn't the influence of all they represent bring Auschwitz that much nearer?'

I thought at the time, and I still think, that the correct answer was not to defend the indefensible. It was no use saying that Yeats, according to friends whose judgment I trust, was a man of singular magnanimity of character, as well as a great poet. It was no use denying the facts, which are broadly true. The honest answer was that there is, in fact, a connection, which literary persons were culpably slow to see, between some kinds of early twentieth-century art and the most imbecile expressions of anti-social feeling.[3] That was one reason, among many, why some of us turned our backs on the art and tried to hack out a new or different way for ourselves.[4]

But though many of those writers dominated

8

literary sensibility for a generation, that is no longer so, or at least to nothing like the same extent. Literature changes more slowly than science. It hasn't the same automatic corrective, and so its misguided periods are longer. But it is ill-considered of scientists to judge writers on the evidence of the period 1914–50.

Those are two of the misunderstandings between the two cultures. I should say, since I began to talk about them—the two cultures, that is—I have had some criticism. Most of my scientific acquaintances think that there is something in it, and so do most of the practising artists I know. But I have been argued with by non-scientists of strong down-to-earth interests. Their view is that it is an over-simplification, and that if one is going to talk in these terms there ought to be at least three cultures. They argue that, though they are not scientists themselves, they would share a good deal of the scientific feeling. They would have as little use—perhaps, since they knew more about it, even less use—for the recent literary culture as the scientists themselves. J. H. Plumb, Alan Bullock and some of my American sociological friends have said that they vigorously refuse to be corralled in a cultural box with people they wouldn't be seen dead with, or to be regarded as helping to produce a climate which would not permit of social hope.

I respect those arguments. The number 2 is a very dangerous number: that is why the dialectic

is a dangerous process. Attempts to divide anything into two ought to be regarded with much
suspicion. I have thought a long time about going
in for further refinements: but in the end I have
decided against. I was searching for something a
little more than a dashing metaphor, a good deal
less than a cultural map: and for those purposes
the two cultures is about right, and subtilising any
more would bring more disadvantages than it's
worth.

At one pole, the scientific culture really is a culture, not only in an intellectual but also in an an-
thropological sense. That is, its members need not,
and of course often do not, always completely understand each other; biologists more often than
not will have a pretty hazy idea of contemporary
physics; but there are common attitudes, common
standards and patterns of behaviour, common ap-
proaches and assumptions. This goes surprisingly
wide and deep. It cuts across other mental patterns, such as those of religion or politics or class.

Statistically, I suppose slightly more scientists
are in religious terms unbelievers, compared with
the rest of the intellectual world—though there are
plenty who are religious, and that seems to be increasingly so among the young. Statistically also,
slightly more scientists are on the Left in open
politics—though again, plenty always have called
themselves conservatives, and that also seems to be
more common among the young. Compared with

the rest of the intellectual world, considerably more scientists in this country and probably in the U.S. come from poor families.[5] Yet, over a whole range of thought and behaviour, none of that matters very much. In their working, and in much of their emotional life, their attitudes are closer to other scientists than to non-scientists who in religion or politics or class have the same labels as themselves. If I were to risk a piece of shorthand, I should say that naturally they had the future in their bones.

They may or may not like it, but they have it. That was as true of the conservatives J. J. Thomson and Lindemann as of the radicals Einstein or Blackett: as true of the Christian A. H. Compton as of the materialist Bernal: of the aristocrats Broglie or Russell as of the proletarian Faraday: of those born rich, like Thomas Merton or Victor Rothschild, as of Rutherford, who was the son of an odd-job handyman. Without thinking about it, they respond alike. That is what a culture means.

At the other pole, the spread of attitudes is wider. It is obvious that between the two, as one moves through intellectual society from the physicists to the literary intellectuals, there are all kinds of tones of feeling on the way. But I believe the pole of total incomprehension of science radiates its influence on all the rest. That total incomprehension gives, much more pervasively than we realise, living in it, an unscientific flavour to the whole 'traditional' culture, and that unscientific flavour is often, much more than we admit, on the point of turning anti-scientific. The feelings of one pole become the anti-feelings of the other. If the scientists have the future in their bones, then the traditional culture responds by wishing the future did not exist.[6] It is the traditional culture, to an extent remarkably little diminished by the emergence of the scientific one, which manages the western world.

This polarisation is sheer loss to us all. To us as people, and to our society. It is at the same time practical and intellectual and creative loss, and I repeat that it is false to imagine that those three considerations are clearly separable. But for a moment I want to concentrate on the intellectual loss.

The degree of incomprehension on both sides is the kind of joke which has gone sour. There are about fifty thousand working scientists in the country and about eighty thousand professional engineers or applied scientists. During the war and in the years since, my colleagues and I have had to interview somewhere between thirty to forty thousand of these—that is, about 25 per cent. The number is large enough to give us a fair sample, though of the men we talked to most would still be under forty. We were able to find out a certain amount of what they read and thought about. I confess that even I, who am fond of them and respect them, was a bit shaken. We hadn't quite expected that the links with the traditional culture should be so tenuous, nothing more than a formal touch of the cap.

As one would expect, some of the very best scientists had and have plenty of energy and interest to spare, and we came across several who had read everything that literary people talk about. But that's very rare. Most of the rest, when one tried to probe for what books they had read, would modestly confess, 'Well, I've *tried* a bit of Dickens', rather as though Dickens were an extraordinarily esoteric, tangled and dubiously rewarding writer, something like Rainer Maria Rilke. In fact that is exactly how they do regard him: we thought that discovery, that Dickens had been transformed into the type-specimen of literary incomprehensibility, was one of the oddest results of the whole exercise.

But of course, in reading him, in reading almost any writer whom we should value, they are just touching their caps to the traditional culture. They have their own culture, intensive, rigorous, and constantly in action. This culture contains a great deal of argument, usually much more rigorous, and almost always at a higher conceptual level, than literary persons' arguments—even though the scientists do cheerfully use words in senses which literary persons don't recognise, the senses are exact ones, and when they talk about 'subjective', 'objective', 'philosophy' or 'progressive',[7] they know what they mean, even though it isn't what one is accustomed to expect.

Remember, these are very intelligent men. Their culture is in many ways an exacting and admirable one. It doesn't contain much art, with the exception, an important exception, of music. Verbal exchange, insistent argument. Long-playing records. Colour-photography. The ear, to some extent the eye. Books, very little, though perhaps not many would go so far as one hero, who perhaps I should admit was further down the scientific ladder than the people I've been talking about—who, when asked what books he read, replied firmly and confidently: 'Books? I prefer to use my books as tools.' It was very hard not to let the mind wander —what sort of tool would a book make? Perhaps a hammer? A primitive digging instrument?

Of books, though, very little. And of the books which to most literary persons are bread and butter, novels, history, poetry, plays, almost nothing at all. It isn't that they're not interested in the psychological or moral or social life. In the social life, they certainly are, more than most of us. In the moral, they are by and large the soundest group of intellectuals we have; there is a moral component right in the grain of science itself, and almost all scientists form their own judgments of the moral life. In the psychological they have as much interest as most of us, though occasionally I fancy they come to it rather late. It isn't that they lack the in-

14

terests. It is much more that the whole literature of the traditional culture doesn't seem to them relevant to those interests. They are, of course, dead wrong. As a result, their imaginative understanding is less than it could be. They are self-impoverished.

But what about the other side? They are impoverished too—perhaps more seriously, because they are vainer about it. They still like to pretend that the traditional culture is the whole of 'culture', as though the natural order didn't exist. As though the exploration of the natural order was of no interest either in its own value or its consequences. As though the scientific edifice of the physical world was not, in its intellectual depth, complexity and articulation, the most beautiful and wonderful collective work of the mind of man. Yet most non-scientists have no conception of that edifice at all. Even if they want to have it, they can't. It is rather as though, over an immense range of intellectual experience, a whole group was tone-deaf. Except that this tone-deafness doesn't come by nature, but by training, or rather the absence of training.

As with the tone-deaf, they don't know what they miss. They give a pitying chuckle at the news of scientists who have never read a major work of English literature. They dismiss them as ignorant specialists. Yet their own ignorance and their own specialisation is just as startling. A good many

15

times I have been present at gatherings of people who, by the standards of the traditional culture, are thought highly educated and who have with considerable gusto been expressing their incredulity at the illiteracy of scientists. Once or twice I have been provoked and have asked the company how many of them could describe the Second Law of Thermodynamics. The response was cold: it was also negative. Yet I was asking something which is about the scientific equivalent of: *Have you read a work of Shakespeare's?*

I now believe that if I had asked an even simpler question—such as, What do you mean by mass, or acceleration, which is the scientific equivalent of saying, *Can you read?*—not more than one in ten of the highly educated would have felt that I was speaking the same language. So the great edifice of modern physics goes up, and the majority of the cleverest people in the western world have about as much insight into it as their neolithic ancestors would have had.

Just one more of those questions, that my non-scientific friends regard as being in the worst of taste. Cambridge is a university where scientists and non-scientists meet every night at dinner.[8] About two years ago, one of the most astonishing experiments in the whole history of science was brought off. I don't mean the sputnik—that was admirable for quite different reasons, as a feat of organisation and a triumphant use of existing

16

knowledge. No, I mean the experiment at Columbia by Yang and Lee. It is an experiment of the greatest beauty and originality, but the result is so startling that one forgets how beautiful the experiment is. It makes us think again about some of the fundamentals of the physical world. Intuition, common sense—they are neatly stood on their heads. The result is usually known as the contradiction of parity. If there were any serious communication between the two cultures, this experiment would have been talked about at every High Table in Cambridge. Was it? I wasn't here: but I should like to ask the question.

There seems then to be no place where the cultures meet. I am not going to waste time saying that this is a pity. It is much worse than that. Soon I shall come to some practical consequences. But at the heart of thought and creation we are letting some of our best chances go by default. The clashing point of two subjects, two disciplines, two cultures—of two galaxies, so far as that goes—ought to produce creative chances. In the history of mental activity that has been where some of the breakthroughs came. The chances are there now. But they are there, as it were, in a vacuum, because those in the two cultures can't talk to each other. It is bizarre how very little of twentieth-century science has been assimilated into twentieth-century art. Now and then one used to find poets conscientiously using scientific expressions, and getting

17

them wrong—there was a time when 'refraction' kept cropping up in verse in a mystifying fashion, and when 'polarised light' was used as though writers were under the illusion that it was a specially admirable kind of light.

Of course, that isn't the way that science could be any good to art. It has got to be assimilated along with, and as part and parcel of, the whole of our mental experience, and used as naturally as the rest.

I said earlier that this cultural divide is not just an English phenomenon: it exists all over the western world. But it probably seems at its sharpest in England, for two reasons. One is our fanatical belief in educational specialisation, which is much more deeply ingrained in us than in any country in the world, west or east. The other is our tendency to let our social forms crystallise. This tendency appears to get stronger, not weaker, the more we iron out economic inequalities: and this is specially true in education. It means that once anything like a cultural divide gets established, all the social forces operate to make it not less rigid, but more so.

The two cultures were already dangerously separate sixty years ago; but a prime minister like Lord Salisbury could have his own laboratory at Hatfield, and Arthur Balfour had a somewhat more than amateur interest in natural science. John Anderson did some research in organic chemistry in Würzburg before passing first into the Civil Service, and incidentally took a spread of subjects which is now impossible.[9] None of that degree of interchange at the top of the Establishment is likely, or indeed thinkable, now.[10]

In fact, the separation between the scientists and non-scientists is much less bridgeable among the young than it was even thirty years ago. Thirty years ago the cultures had long ceased to speak to each other: but at least they managed a kind of frozen smile across the gulf. Now the politeness has gone, and they just make faces. It is not only that the young scientists now feel that they are part of a culture on the rise while the other is in retreat. It is also, to be brutal, that the young scientists know that with an indifferent degree they'll get a comfortable job, while their contemporaries and counterparts in English or History will be lucky to earn 60 per cent as much. No young scientist of any talent would feel that he isn't wanted or that his work is ridiculous, as did the hero of *Lucky Jim*, and in fact, some of the disgruntlement of Amis and his associates is the disgruntlement of the under-employed arts graduate.

There is only one way out of all this: it is, of course, by rethinking our education. In this country, for the two reasons I have given, that is more difficult than in any other. Nearly everyone will agree that our school education is too specialised. But nearly everyone feels that it is outside the will

19

of man to alter it. Other countries are as dissatisfied with their education as we are, but are not so
resigned.

The U.S. teach out of proportion more children
up to eighteen than we do: they teach them far
more widely, but nothing like so rigorously. They
know that: they are hoping to take the problem in
hand within ten years, though they may not have
all that time to spare. The U.S.S.R. also teach out
of proportion more children than we do: they also
teach far more widely than we do (it is an absurd
western myth that their school education is specialised) but much too rigorously.[11] They know that—
and they are beating about to get it right. The
Scandinavians, in particular the Swedes, who
would make a more sensible job of it than any of
us, are handicapped by their practical need to devote an inordinate amount of time to foreign lan-
guages. But they too are seized of the problem.

Are we? Have we crystallised so far that we are
no longer flexible at all?

Talk to schoolmasters, and they say that our in-
tense specialisation, like nothing else on earth, is
dictated by the Oxford and Cambridge scholarship
examinations. If that is so, one would have
thought it not utterly impracticable to change the
Oxford and Cambridge scholarship examinations.
Yet one would underestimate the national capacity
for the intricate defensive to believe that that was
easy. All the lessons of our educational history sug-

20

gest we are only capable of increasing specialisation, not decreasing it.

Somehow we have set ourselves the task of producing a tiny *élite*—far smaller proportionately than in any comparable country—educated in one academic skill. For a hundred and fifty years in Cambridge it was mathematics: then it was mathematics or classics: then natural science was allowed in. But still the choice had to be a single one.

It may well be that this process has gone too far to be reversible. I have given reasons why I think it is a disastrous process, for the purpose of a living culture. I am going on to give reasons why I think it is fatal, if we're to perform our practical tasks in the world. But I can think of only one example, in the whole of English educational history, where our pursuit of specialised mental exercises was resisted with success.

It was done here in Cambridge, fifty years ago, when the old order-of-merit in the Mathematical Tripos was abolished. For over a hundred years, the nature of the Tripos had been crystallising. The competition for the top places had got fiercer, and careers hung on them. In most colleges, certainly in my own, if one managed to come out as Senior or Second Wrangler, one was elected a Fellow out of hand. A whole apparatus of coaching had grown up. Men of the quality of Hardy, Littlewood, Russell, Eddington, Jeans, Keynes, went in for two or three years' training for an examination which was intensely competitive and intensely difficult. Most people in Cambridge were very proud of it, with a similar pride to that which almost anyone in England always has for our existing educational institutions, whatever they happen to be. If you study the fly-sheets of the time, you will find the passionate arguments for keeping the examination precisely as it was to all eternity: it was the only way to keep up standards, it was the only fair test of merit, indeed, the only seriously objective test in the world. The arguments, in fact, were almost exactly those which are used today with precisely the same passionate sincerity if anyone suggests that the scholarship examinations might conceivably not be immune from change.

In every respect but one, in fact, the old Mathematical Tripos seemed perfect. The one exception, however, appeared to some to be rather important. It was simply—so the young creative mathematicians, such as Hardy and Littlewood, kept saying —that the training had no intellectual merit at all. They went a little further, and said that the Tripos had killed serious mathematics in England stone dead for a hundred years. Well, even in academic controversy, that took some skirting round, and they got their way. But I have an impression that Cambridge was a good deal more flexible between 1850 and 1914 than it has been in our time. If we had had the old Mathematical Tripos firmly planted among us, should we have ever managed to abolish it?

II

INTELLECTUALS

AS NATURAL LUDDITES

The reasons for the existence of the two cultures are many, deep, and complex, some rooted in social histories, some in personal histories, and some in the inner dynamic of the different kinds of mental activity themselves. But I want to isolate one which is not so much a reason as a correlative, something which winds in and out of any of these discussions. It can be said simply, and it is this. If we forget the scientific culture, then the rest of western intellectuals have never tried, wanted, or been able to understand the industrial revolution, much less accept it. Intellectuals, in particular literary intellectuals, are natural Luddites.

That is specially true of this country, where the industrial revolution happened to us earlier than

23

elsewhere, during a long spell of absentminded-
ness. Perhaps that helps explain our present
degree of crystallisation. But, with a little quali-
fication, it is also true, and surprisingly true, of the
United States.

In both countries, and indeed all over the West,
the first wave of the industrial revolution crept on,
without anyone noticing what was happening. It
was, of course—or at least it was destined to become, under our own eyes, and in our own time—
by far the biggest transformation in society since
the discovery of agriculture. In fact, those two
revolutions, the agricultural and the industrial-
scientific, are the only qualitative changes in social
living that men have ever known. But the traditional culture didn't notice: or when it did notice,
didn't like what it saw. Not that the traditional
culture wasn't doing extremely well out of the revolution; the English educational institutions took
their slice of the English nineteenth-century
wealth, and perversely, it helped crystallise them
in the forms we know.

Almost none of the talent, almost none of the
imaginative energy, went back into the revolution
which was producing the wealth. The traditional
culture became more abstracted from it as it became more wealthy, trained its young men for ad-
ministration, for the Indian Empire, for the
purpose of perpetuating the culture itself, but
never in any circumstances to equip them to un-

24

derstand the revolution or take part in it. Far-sighted men were beginning to see, before the middle of the nineteenth century, that in order to go on producing wealth, the country needed to train some of its bright minds in science, particularly in applied science. No one listened. The traditional culture didn't listen at all: and the pure scientists, such as there were, didn't listen very eagerly. You will find the story, which in spirit continues down to the present day, in Eric Ashby's *Technology and the Academics*.[12]

The academics had nothing to do with the industrial revolution; as Corrie, the old Master of Jesus, said about trains running into Cambridge on Sunday, 'It is equally displeasing to God and to myself'. So far as there was any thinking in nineteenth-century industry, it was left to cranks and clever workmen. American social historians have told me that much the same was true of the U.S. The industrial revolution, which began developing in New England fifty years or so later than ours,[13] apparently received very little educated talent, either then or later in the nineteenth century. It had to make do with the guidance handymen could give it—sometimes, of course, handymen like Henry Ford, with a dash of genius.

The curious thing was that in Germany, in the 1830's and 1840's, long before serious industrialisation had started there, it was possible to get a good university education in applied science, bet·

ter than anything England or the U.S. could offer for a couple of generations. I don't begin to understand this: it doesn't make social sense: but it was so. With the result that Ludwig Mond, the son of a court purveyor, went to Heidelberg and learnt some sound applied chemistry. Siemens, a Prussian signals officer, at military academy and university went through what for their time were excellent courses in electrical engineering. Then they came to England, met no competition at all, brought in other educated Germans, and made fortunes exactly as though they were dealing with a rich, illiterate colonial territory. Similar fortunes were made by German technologists in the United States.

Almost everywhere, though, intellectual persons didn't comprehend what was happening. Certainly the writers didn't. Plenty of them shuddered away, as though the right course for a man of feeling was to contract out; some, like Ruskin and William Morris and Thoreau and Emerson and Lawrence, tried various kinds of fancies which were not in effect more than screams of horror. It is hard to think of a writer of high class who really stretched his imaginative sympathy, who could see at once the hideous back-streets, the smoking chimneys, the internal price—and also the prospects of life that were opening out for the poor, the intimations, up to now unknown except to the lucky, which were just coming within reach of the remaining 99 per

26

cent of his brother men. Some of the nineteenth-century Russian novelists might have done; their natures were broad enough; but they were living in a pre-industrial society and didn't have the opportunity. The only writer of world class who seems to have had an understanding of the industrial revolution was Ibsen in his old age: and there wasn't much that old man didn't understand.

For, of course, one truth is straightforward. Industrialisation is the only hope of the poor. I use the word 'hope' in a crude and prosaic sense. I have not much use for the moral sensibility of anyone who is too refined to use it so. It is all very well for us, sitting pretty, to think that material standards of living don't matter all that much. It is all very well for one, as a personal choice, to reject industrialisation—do a modern Walden, if you like, and if you go without much food, see most of your children die in infancy, despise the comforts of literacy, accept twenty years off your own life, then I respect you for the strength of your aesthetic revulsion.[14] But I don't respect you in the slightest if, even passively, you try to impose the same choice on others who are not free to choose. In fact, we know what their choice would be. For, with singular unanimity, in any country where they have had the chance, the poor have walked off the land into the factories as fast as the factories could take them.

I remember talking to my grandfather when I

27

was a child. He was a good specimen of a nine-teenth-century artisan. He was highly intelligent, and he had a great deal of character. He had left school at the age of ten, and had educated himself intensely until he was an old man. He had all his class's passionate faith in education. Yet, he had never had the luck—or, as I now suspect, the worldly force and dexterity—to go very far. In fact, he never went further than maintenance fore-man in a tramway depot. His life would seem to his grandchildren laborious and unrewarding al-most beyond belief. But it didn't seem to him quite like that. He was much too sensible a man not to know that he hadn't been adequately used: he had too much pride not to feel a proper ran-cour: he was disappointed that he had not done more—and yet, compared with *his* grandfather, he felt he had done a lot. His grandfather must have been an agricultural labourer. I don't so much as know his Christian name. He was one of the 'dark people', as the old Russian liberals used to call them, completely lost in the great anonymous sludge of history. So far as my grandfather knew, he could not read or write. He was a man of ability, my grandfather thought; my grandfather was pretty unforgiving about what society had done, or not done, to his ancestors, and did not romanticise their state. It was no fun being an agricultural labourer in the mid- to late eight-eenth century, in the time that we, snobs that we

28

are, think of only as the time of the Enlightenment and Jane Austen.

The industrial revolution looked very different according to whether one saw it from above or below. It looks very different today according to whether one sees it from Chelsea or from a village in Asia. To people like my grandfather, there was no question that the industrial revolution was less bad than what had gone before. The only question was, how to make it better.

In a more sophisticated sense, that is still the question. In the advanced countries, we have realised in a rough and ready way what the old industrial revolution brought with it. A great increase of population, because applied science went hand in hand with medical science and medical care. Enough to eat, for a similar reason. Everyone able to read and write, because an industrial society can't work without. Health, food, education; nothing but the industrial revolution could have spread them right down to the very poor. Those are primary gains—there are losses [15] too, of course, one of which is that organising a society for industry makes it easy to organise it for all-out war. But the gains remain. They are the base of our social hope.

And yet: do we understand how they have happened? Have we begun to comprehend even the old industrial revolution? Much less the new scientific revolution in which we stand? There never was anything more necessary to comprehend.

III

THE SCIENTIFIC

REVOLUTION

I have just mentioned a distinction between the industrial revolution and the scientific revolution. The distinction is not clear-edged, but it is a useful one, and I ought to try to define it now. By the industrial revolution, I mean the gradual use of machines, the employment of men and women in factories, the change in this country from a population mainly of agricultural labourers to a population mainly engaged in making things in factories and distributing them when they were made. That change, as I have said, crept on us unawares, untouched by academics, hated by Luddites, practical Luddites and intellectual ones. It is connected, so it seems to me, with many of the attitudes to science and aesthetics which have crys-

30

tallised among us. One can date it roughly from the middle of the eighteenth century to the early twentieth. Out of it grew another change, closely related to the first, but far more deeply scientific, far quicker, and probably far more prodigious in its result. This change comes from the application of real science to industry, no longer hit and miss, no longer the ideas of odd 'inventors', but the real stuff.

Dating this second change is very largely a matter of taste. Some would prefer to go back to the first large-scale chemical or engineering industries, round about sixty years ago. For myself, I should put it much further on, not earlier than thirty to forty years ago—and as a rough definition, I should take the time when atomic particles were first made industrial use of. I believe the industrial society of electronics, atomic energy, automation, is in cardinal respects different in kind from any that has gone before, and will change the world much more. It is this transformation that, in my view, is entitled to the name of 'scientific revolution'.

This is the material basis of our lives: or more exactly, the social plasma of which we are a part. And we know almost nothing about it. I remarked earlier that highly educated members of the non-scientific culture couldn't cope with the simplest concepts of pure science: it is unexpected, but they would be even less happy with applied science.

How many educated people know anything about productive industry, old-style or new? What is a machine-tool? I once asked a literary party; and they looked shifty. Unless one knows, industrial production is as mysterious as witch-doctoring. Or take buttons. Buttons aren't very complicated things: they are being made in millions every day: one has to be a reasonably ferocious Luddite not to think that that is, on the whole, an estimable activity. Yet I would bet that out of men getting firsts in arts subjects at Cambridge this year, not one in ten could give the loosest analysis of the human organisation which it needs.

In the United States, perhaps, there is a wider nodding acquaintance with industry, but, now I come to think of it, no American novelist of any class has ever been able to assume that his audience had it. He can assume, and only too often does, an acquaintance with a pseudo-feudal society, like the fag-end of the Old South—but not with industrial society. Certainly an English novelist couldn't.

Yet the personal relations in a productive organisation are of the greatest subtlety and interest. They are very deceptive. They look as though they ought to be the personal relations that one gets in any hierarchical structure with a chain of command, like a division in the army or a department in the civil service. In practice they are much more complex than that, and anyone used to the straight chain of command gets lost the instant he sets foot in an industrial organisation. No one in any country, incidentally, knows yet what these personal relations ought to be. That is a problem almost independent of large-scale politics, a problem springing straight out of the industrial life.

I think it is only fair to say that most pure scientists have themselves been devastatingly ignorant of productive industry, and many still are. It is permissible to lump pure and applied scientists into the same scientific culture, but the gaps are wide. Pure scientists and engineers often totally misunderstand each other. Their behaviour tends to be very different: engineers have to live their lives in an organised community, and however odd they are underneath they manage to present a disciplined face to the world. Not so pure scientists. In the same way pure scientists still, though less than twenty years ago, have statistically a higher proportion in politics left of centre than any other profession: not so engineers, who are conservative almost to a man. Not reactionary in the extreme literary sense, but just conservative. They are absorbed in making things, and the present social order is good enough for them.

Pure scientists have by and large been dimwitted about engineers and applied science. They couldn't get interested. They wouldn't recognise that many of the problems were as intellectually exacting as pure problems, and that many of the solutions were as satisfying and beautiful. Their instinct—perhaps sharpened in this country by the passion to find a new snobbism wherever possible, and to invent one if it doesn't exist—was to take it for granted that applied science was an occupation for second-rate minds. I say this more sharply because thirty years ago I took precisely that line myself. The climate of thought of young research workers in Cambridge then was not to our credit. We prided ourselves that the science we were doing could not, in any conceivable circumstances, have any practical use. The more firmly one could make that claim, the more superior one felt.

Rutherford himself had little feeling for engineering. He was amazed—he used to relate the story with incredulous admiration—that Kapitza had actually sent an engineering drawing to Metrovick, and that those magicians had duly studied the drawing, *made the machine,* and delivered it in Kapitza's laboratory! Rutherford was so impressed by Cockcroft's engineering skill that he secured for him a special capital grant for machinery—the grant was as much as six hundred pounds! In 1933, four years before his death, Rutherford said, firmly and explicitly, that he didn't believe the energy of the nucleus would ever be released—nine years later, at Chicago, the first pile began to run. That was the only major bloomer in scientific judgment Rutherford ever made. It is interesting that it should be at the point where pure science turned into applied.

No, pure scientists did not show much understanding or display much sense of social fact. The best that can be said for them is that, given the necessity, they found it fairly easy to learn. In the war, a great many scientists had to learn, for the good Johnsonian reason that sharpens one's wits, something about productive industry. It opened their eyes. In my own job, I had to try to get some insight into industry. It was one of the most valuable pieces of education in my life. But it started when I was thirty-five, and I ought to have had it much earlier.

That brings me back to education. Why aren't we coping with the scientific revolution? Why are other countries doing better? How are we going to meet our future, both our cultural and practical future? It should be obvious by now that I believe both lines of argument lead to the same end. If one begins by thinking only of the intellectual life, or only of the social life, one comes to a point where it becomes manifest that our education has gone wrong, and gone wrong in the same way.

I don't pretend that any country has got its education perfect. In some ways, as I said before, the Russians and Americans are both more actively dissatisfied with theirs than we are: that is, they are taking more drastic steps to change it. But that is because they are more senstive to the world they are living in. For myself, I have no doubt that, though neither of them have got the answer right, they are a good deal nearer than we are. We do some things much better than either of them. In educational tactics, we are often more gifted than they are. In educational strategy, by their side we are only playing at it.

The differences between the three systems are revelatory. We teach, of course, a far smaller proportion of our children up to the age of eighteen: and we take a far smaller proportion even of those we do teach up to the level of a university degree. The old pattern of training a small *élite* has never been broken, though it has been slightly bent. Within that pattern, we have kept the national passion for specialisation: and we work our clever young up to the age of twenty-one far harder than the Americans, though no harder than the Russians. At eighteen, our science specialists know more science than their contemporaries anywhere, though they know less of anything else. At twenty-one, when they take their first degree, they are probably still a year or so ahead.

The American strategy is different in kind. They take everyone, the entire population,[16] up to eighteen in high schools, and educate them very loosely and generally. Their problem is to inject some rigour—in particular some fundamental mathematics and science—into this loose education. A very large proportion of the eighteen-year-olds then go to college: and this college education is, like the school education, much more diffuse and

36

less professional than ours.[17] At the end of four years, the young men and women are usually not so well-trained professionally as we are: though I think it is fair comment to say that a higher proportion of the best of them, having been run on a looser rein, retain their creative zest. Real severity enters with the Ph.D. At that level the Americans suddenly begin to work their students much harder than we do. It is worth remembering that they find enough talent to turn out nearly as many Ph.D.'s in science and engineering each year as we contrive to get through our first degrees.

The Russian high school education is much less specialised than ours, much more arduous than the American. It is so arduous that for the non-academic it seems to have proved too tough, and they are trying other methods from fifteen to seventeen. The general method has been to put everyone through a kind of continental Lycée course, with a sizeable component, more than 40 per cent, of science and mathematics. Everyone has to do all subjects. At the university this general education ceases abruptly: and for the last three years of the five-year course the specialisation is more intensive even than ours. That is, at most English universities a young man can take an honours degree in mechanical engineering. In Russia he can take, and an enormous number do take, a corresponding degree in one bit of mechanical engineering, as it might be aerodynamics or machine-tool design or diesel engine production.

They won't listen to me, but I believe they have overdone this, just as I believe they have slightly overdone the number of engineers they are training. It is now much larger than the rest of the world put together—getting on for fifty per cent larger.[18] Pure scientists they are training only slightly more than the United States, though in physics and mathematics the balance is heavily in the Russian direction.

Our population is small by the side of either the U.S.A. or the U.S.S.R. Roughly, if we compare like with like, and put scientists and engineers together, we are training at a professional level per head of the population one Englishman to every one and a half Americans to every two and a half Russians. [19] Someone is wrong.

With some qualifications, I believe the Russians have judged the situation sensibly. They have a deeper insight into the scientific revolution than we have, or than the Americans have. The gap between the cultures doesn't seem to be anything like so wide as with us. If one reads contemporary Soviet novels, for example, one finds that their novelists can assume in their audience—as we cannot —at least a rudimentary acquaintance with what industry is all about. Pure science doesn't often come in, and they don't appear much happier with it than literary intellectuals are here. But engineering does come in. An engineer in a Soviet novel is as acceptable, so it seems, as a psychiatrist in an American one. They are as ready to cope in art with the process of production as Balzac was with the processes of craft manufacture. I don't want to overstress this, but it may be significant. It may also be significant that, in these novels, one is constantly coming up against a passionate belief in education. The people in them believe in education exactly as my grandfather did, and for the same mixture of idealistic and bread-and-butter reasons.

Anyway, the Russians have judged what kind and number of educated men and women [20] a country needs to come out top in the scientific revolution. I am going to oversimplify, but their estimate, and I believe it's pretty near right, is this. First of all, as many alpha plus scientists as the country can throw up. No country has many of them. Provided the schools and universities are there, it doesn't matter all that much what you teach them. They will look after themselves.[21] We probably have at least as many pro-rata as the Russians and Americans; that is the least of our worries. Second, a much larger stratum of alpha professionals—these are the people who are going to do the supporting research, the high class design and development. In quality, England compares well in this stratum with the U.S.A. or U.S.S.R.: this is what our education is specially

39

geared to produce. In quantity, though, we are not discovering (again per head of the population) half as many as the Russians think necessary and are able to find. Third, another stratum, educated to about the level of Part I of the Natural Sciences or Mechanical Sciences Tripos, or perhaps slightly below that. Some of these will do the secondary technical jobs, but some will take major responsibility, particularly in the human jobs. The proper use of such men depends upon a different distribution of ability from the one that has grown up here. As the scientific revolution goes on, the call for these men will be something we haven't imagined, though the Russians have. They will be required in thousands upon thousands, and they will need all the human development that university education can give them.[22] It is here, perhaps, most of all that our insight has been fogged. Fourthly and last, politicians, administrators, an entire community, who know enough science to have a sense of what the scientists are talking about.

That, or something like that, is the specification for the scientific revolution.[23] I wish I were certain that in this country we were adaptable enough to meet it. In a moment I want to go on to an issue which will, in the world view, count more: but perhaps I can be forgiven for taking a sideways look at our own fate. It happens that of all the advanced countries, our position is by a long way the most precarious. That is the result of history and accident, and isn't to be laid to the blame of any Englishman now living. If our ancestors had invested talent in the industrial revolution instead of the Indian Empire, we might be more soundly based now. But they didn't.

We are left with a population twice as large as we can grow food for, so that we are always going to be *au fond* more anxious than France or Sweden: [24] and with very little in the way of natural resources—by the standard of the great world powers, with nothing. The only real assets we have, in fact, are our wits. Those have served us pretty well, in two ways. We have a good deal of cunning, native or acquired, in the arts of getting on among ourselves: that is a strength. And we have been inventive and creative, possibly out of proportion to our numbers. I don't believe much in national differences in cleverness, but compared with other countries we are certainly no stupider.

Given these two assets, and they are our only ones, it should have been for us to understand the scientific revolution first, to educate ourselves to the limit, and give a lead. Well, we have done something. In some fields, like atomic energy, we have done better than anyone could have predicted. Within the pattern, the rigid and crystallised pattern of our education and of the two cultures, we have been trying moderately hard to adjust ourselves.

41

The bitterness is, it is nothing like enough. To say we have to educate ourselves or perish, is a little more melodramatic than the facts warrant. To say, we have to educate ourselves or watch a steep decline in our own lifetime, is about right. We can't do it, I am now convinced, without breaking the existing pattern. I know how difficult this is. It goes against the emotional grain of nearly all of us. In many ways, it goes against my own, standing uneasily with one foot in a dead or dying world and the other in a world that at all costs we must see born. I wish I could be certain that we shall have the courage of what our minds tell us.

More often than I like, I am saddened by a historical myth. Whether the myth is good history or not, doesn't matter; it is pressing enough for me. I can't help thinking of the Venetian Republic in their last half-century. Like us, they had once been fabulously lucky. They had become rich, as we did, by accident. They had acquired immense political skill, just as we have. A good many of them were tough-minded, realistic, patriotic men. They knew, just as clearly as we know, that the current of history had begun to flow against them. Many of them gave their minds to working out ways to keep going. It would have meant breaking the pattern into which they had crystallised. They were fond of the pattern, just as we are fond of ours. They never found the will to break it.

IV

THE RICH AND THE POOR

But that is our local problem, and it is for us to struggle with it. Sometimes, it is true, I have felt that the Venetian shadow falls over the entire West. I have felt that on the other side of the Mississippi. In more resilient moments, I comfort myself that Americans are much more like us between 1850 and 1914. Whatever they don't do, they do react. It's going to take them a long and violent pull to be as well prepared for the scientific revolution as the Russians are, but there are good chances that they will do it.

Nevertheless, that isn't the main issue of the scientific revolution. The main issue is that the people in the industrialised countries are getting richer, and those in the non-industrialised countries are at best standing still: so that the gap between the industrialised countries and the rest is widening every day. On the world scale this is the gap between the rich and the poor.

Among the rich are the U.S., the white Commonwealth countries, Great Britain, most of Europe, and the U.S.S.R. China is betwixt and between, not yet over the industrial hump, but probably getting there. The poor are all the rest. In the rich countries people are living longer, eating better, working less. In a poor country like India, the expectation of life is less than half what it is in England. There is some evidence that Indians and other Asians are eating less, in absolute quantities, than they were a generation ago. The statistics are not reliable, and informants in the F.A.O. have told me not to put much trust in them. But it is accepted that, in all non-industrialised countries, people are not eating better than at the subsistence level. And they are working as people have always had to work, from Neolithic times until our own. Life for the overwhelming majority of mankind has always been nasty, brutish and short. It is so in the poor countries still.

This disparity between the rich and the poor has been noticed. It has been noticed, most acutely and not unnaturally, by the poor. Just because they have noticed it, it won't last for long. Whatever else in the world we know survives to the year 2000, that won't. Once the trick of getting rich is known, as it now is, the world can't survive half rich and half poor. It's just not on.

The West has got to help in this transformation. The trouble is, the West with its divided culture finds it hard to grasp just how big, and above all just how fast, the transformation must be.

Earlier I said that few non-scientists really understand the scientific concept of acceleration. I meant that as a gibe. But in social terms, it is a little more than a gibe. During all human history until this century, the rate of social change has been very slow. So slow, that it would pass unnoticed in one person's lifetime. That is no longer so. The rate of change has increased so much that our imagination can't keep up. There is *bound* to be more social change, affecting more people, in the next decade than in any before. There is *bound* to be more change again, in the 1970's. In the poor countries, people have caught on to this simple concept. Men there are no longer prepared to wait for periods longer than one person's lifetime.

The comforting assurances, given *de haut en bas*, that maybe in a hundred or two hundred years things may be slightly better for them—they only madden. Pronouncements such as one still hears from old Asia or old Africa hands—Why, it will take those people five hundred years to get up to our standard!—they are both suicidal and technologically illiterate. Particularly when said, as they always seem to be said, by someone looking

45

as though it wouldn't take Neanderthal Man five years to catch up with *him*.

The fact is, the rate of change has already been proved possible. Someone said, when the first atomic bomb went off, that the only important secret is now let out—the thing works. After that, any determined country could make the bomb, given a few years. In the same way, the only secret of the Russian and Chinese industrialisation is that they've brought it off. That is what Asians and Africans have noticed. It took the Russians about forty years, starting with something of an industrial base—Tsarist industry wasn't negligible—but interrupted by a civil war and then the greatest war of all. The Chinese started with much less of an industrial base, but haven't been interrupted, and it looks like taking them not much over half the time.

These transformations were made with inordinate effort and with great suffering. Much of the suffering was unnecessary: the horror is hard to look at straight, standing in the same decades. Yet they've proved that common men can show astonishing fortitude in chasing jam tomorrow. Jam today, and men aren't at their most exciting: jam tomorrow, and one often sees them at their noblest. The transformations have also proved something which only the scientific culture can take in its stride. Yet, when we don't take it in our stride, it makes us look silly.

46

It is simply that technology is rather easy. Or more exactly, technology is the branch of human experience that people can learn with predictable results. For a long time, the West misjudged this very badly. After all, a good many Englishmen have been skilled in mechanical crafts for half-a-dozen generations. Somehow we've made ourselves believe that the whole of technology was a more or less incommunicable art. It's true enough, we start with a certain advantage. Not so much because of tradition, I think, as because all our children play with mechanical toys. They are picking up pieces of applied science before they can read. That is an advantage we haven't made the most of. Just as the Americans have the advantage that nine out of ten adults can drive a car and are to some extent mechanics. In the last war, which was a war of small machines, that was a real military asset. Russia is catching up with the U.S. in major industry— but it will be a long time before Russia is as convenient a country as the U.S. in which to have one's car break down.[25]

The curious thing is, none of that seems to matter much. For the task of totally industrialising a major country, as in China today, it only takes will to train enough scientists and engineers and technicians. Will, and quite a small number of years. There is no evidence that any country or race is better than any other in scientific teachability: there is a good deal of evidence that all are

47

much alike. Tradition and technical background seem to count for surprisingly little.

We've all seen this with our own eyes. I myself have found Sicilian girls taking the top places in the Honours Physics course—a very exacting course —at the University of Rome: they'd have been in something like purdah thirty years ago. And I remember John Cockcroft coming back from Moscow some time in the early 1930's. The news got round that he had been able to have a look, not only at laboratories, but at factories and the mechanics in them. What we expected to hear, I don't know: but there were certainly some who had pleasurable expectations of those stories precious to the hearts of western man, about moujiks prostrating themselves before a milling machine, or breaking a vertical borer with their bare hands. Someone asked Cockcroft what the skilled workmen were like. Well, he has never been a man to waste words. A fact is a fact is a fact. 'Oh,' he said, 'they're just about the same as the ones at Metrovick.' That was all. He was, as usual, right.

There is no getting away from it. It is technically possible to carry out the scientific revolution in India, Africa, South-east Asia, Latin America, the Middle East, within fifty years. There is no excuse for western man not to know this. And not to know that this is the one way out through the three menaces which stand in our way—H-bomb war, over-population, the gap between the rich and the poor. This is one of the situations where the worst crime is innocence.

Since the gap between the rich countries and the poor can be removed, it will be. If we are shortsighted, inept, incapable either of good-will or enlightened self-interest, then it may be removed to the accompaniment of war and starvation: but removed it will be. The questions are, how, and by whom. To those questions, one can only give partial answers; but that may be enough to set us thinking. The scientific revolution on the world-scale needs, first and foremost, capital: capital in all forms, including capital machinery. The poor countries, until they have got beyond a certain point on the industrial curve, cannot accumulate that capital. That is why the gap between rich and poor is widening. The capital must come from outside.

There are only two possible sources. One is the West, which means mainly the U.S., the other is the U.S.S.R. Even the United States hasn't infinite resources of such capital. If they or Russia tried to do it alone, it would mean an effort greater than either had to make industrially in the war. If they both took part, it wouldn't mean that order of sacrifice—though in my view it's optimistic to think, as some wise men do, that it would mean no sacrifice at all. The scale of the operation requires that it would have to be a national one. Private industry, even the biggest private industry, can't touch it, and in no sense is it a fair business risk. It's a bit like asking Duponts or I.C.I. back in 1940 to finance the entire development of the atomic bomb.

The second requirement, after capital, as important as capital, is men. That is, trained scientists and engineers adaptable enough to devote themselves to a foreign country's industrialisation for at least ten years out of their lives. Here, unless and until the Americans and we educate ourselves both sensibly and imaginatively, the Russians have a clear edge. This is where their educational policy has already paid big dividends. They have such men to spare if they are needed. We just haven't, and the Americans aren't much better off. Imagine, for example, that the U.S. government and ours had agreed to help the Indians to carry out a major industrialisation, similar in scale to the Chinese. Imagine that the capital could be found. It would then require something like ten thousand to twenty thousand engineers from the U.S. and here to help get the thing going. At present, we couldn't find them.

These men, whom we don't yet possess, need to be trained not only in scientific but in human terms. They could not do their job if they did not shrug off every trace of paternalism. Plenty of Europeans, from St Francis Xavier to Schweitzer, have devoted their lives to Asians and Africans, nobly but paternally. These are not the Europeans whom Asians and Africans are going to welcome now. They want men who will muck in as colleagues, who will pass on what they know, do an honest technical job, and get out. Fortunately, this is an attitude which comes easily to scientists. They are freer than most people from racial feeling; their own culture is in its human relations a democratic one. In their own internal climate, the breeze of the equality of man hits you in the face, sometimes rather roughly, just as it does in Norway.

That is why scientists would do us good all over Asia and Africa. And they would do their part too in the third essential of the scientific revolution—which, in a country like India, would have to run in parallel with the capital investment and the initial foreign help. That is, an educational programme as complete as the Chinese, who appear in ten years to have transformed their universities and built so many new ones that they are now nearly independent of scientists and engineers from outside. Ten years. With scientific teachers from this country and the U.S., and what is also necessary, with teachers of English, other poor countries could do the same in twenty.

That is the size of the problem. An immense capital outlay, an immense investment in men, both scientists and linguists, most of whom the West does not yet possess. With rewards negligible in the short term, apart from doing the job: and in the long term most uncertain.

People will ask me, in fact in private they have already asked me—'This is all very fine and large. But you are supposed to be a realistic man. You are interested in the fine structure of politics; you have spent some time studying how men behave in the pursuit of their own ends. Can you possibly believe that men will behave as you say they ought to? Can you imagine a political technique, in parliamentary societies like the U.S. or our own, by which any such plan could become real? Do you really believe that there is one chance in ten that any of this will happen?'

That is fair comment. I can only reply that I don't know. On the one hand, it is a mistake, and it is a mistake, of course, which anyone who is called realistic is specially liable to fall into, to think that when we have said something about the egotisms, the weaknesses, the vanities, the power-seekings of men, that we have said everything. Yes, they are like that. They are the bricks with which we have got to build, and one can judge them through the extent of one's own selfishness. But they are sometimes capable of more, and any 'realism' which doesn't admit of that isn't serious.

On the other hand, I confess, and I should be less than honest if I didn't, that I can't see the political techniques through which the good human capabilities of the West can get into action. The best one can do, and it is a poor best, is to nag away. That is perhaps too easy a palliative for one's disquiet. For, though I don't know how we can do what we need to do, or whether we shall do anything at all, I do know this: that, if we don't do it, the Communist countries will in time. They will do it at great cost to themselves and others, but they will do it. If that is how it turns out, we shall have failed, both practically and morally. At best, the West will have become an *enclave* in a different world—and this country will be the *enclave* of an *enclave*. Are we resigning ourselves to that? History is merciless to failure. In any case, if that happens, we shall not be writing the history.

Meanwhile, there are steps to be taken which aren't outside the powers of reflective people. Education isn't the total solution to this problem: but without education the West can't even begin to cope. All the arrows point the same way. Closing the gap between our cultures is a necessity in the most abstract intellectual sense, as well as in the most practical. When those two senses have grown apart, then no society is going to be able to think with wisdom. For the sake of the intellectual life, for the sake of this country's special danger, for the sake of the western society living precariously rich among the poor, for the sake of the poor who needn't be poor if there is intelligence in the world, it is obligatory for us and the Americans and the whole West to look at our education with fresh eyes. This is one of the cases where we and the Americans have the most to learn from each other. We have each a good deal to learn from the Russians, if we are not too proud. Incidentally, the Russians have a good deal to learn from us, too.

Isn't it time we began? The danger is, we have been brought up to think as though we had all the time in the world. We have very little time. So little that I dare not guess at it.

NOTES

1 'The Two Cultures', *New Statesman*, 6 October 1956.

2 This lecture was delivered to a Cambridge audience, and so I used some points of reference which I did not need to explain. G. H. Hardy, 1877–1947, was one of the most distinguished pure mathematicians of his time, and a picturesque figure in Cambridge both as a young don and on his return in 1931 to the Sadleirian Chair of Mathematics.

3 I said a little more about this connection in *The Times Literary Supplement*, 'Challenge to the Intellect', 15 August 1958. I hope some day to carry the analysis further.

4 It would be more accurate to say that, for literary reasons, we felt the prevailing literary modes were useless to us. We were, however, reinforced in that feeling when it occurred to us that those prevailing modes went hand in hand with social attitudes either wicked, or absurd, or both.

5 An analysis of the schools from which Fellows of the Royal Society come tells its own story. The distribution is markedly different from that of, for example, members of the Foreign Service or Queen's Counsel.

6 Compare George Orwell's *1984*, which is the strongest possible wish that the future should not exist, with J. D. Bernal's *World Without War*.

7 *Subjective*, in contemporary technological jargon, means 'divided according to subjects'. *Objective* means 'directed towards an object'. *Philosophy* means 'general intellectual approach or attitude' (for example, a scientist's 'philosophy of guided weapons' might lead him to propose certain kinds of 'objective research'). A 'progressive' job means one with possibilities of promotion.

8 Almost all college High Tables contain Fellows in both scientific and non-scientific subjects.

9 He took the examination in 1905.

10 It is, however, true to say that the compact nature of the managerial layers of English society—the fact that 'everyone knows everyone else'—means that scientists and non-scientists do in fact know each other as people more easily than in most countries. It is also true that a good many leading politicians and administrators keep up lively intellectual and artistic interests to a much greater extent, so far as I can judge, than is the case in the U.S. These are both among our assets.

11 I tried to compare American, Soviet and English education in 'New Minds for the New World', *New Statesman*, 6 September 1956.

12 The best, and almost the only, book on the subject.

13 It developed very fast. An English commission of inquiry into industrial productivity went over to the United States as early as 1865.

14 It is reasonable for intellectuals to prefer to live in the eighteenth-century streets of Stockholm rather than in Vållingby. I should myself. But it is not reasonable for them to obstruct other Vållingbys being built.

15 It is worth remembering that there must have been similar losses—spread over a much longer period—when men changed from the hunting and food gathering life to agriculture. For some, it must have been a genuine spiritual impoverishment.

16 This is not quite exact. In the states where higher education is most completely developed, for example, Wisconsin, about 95 per cent of children attend High School up to eighteen.

17 The U.S. is a complex and plural society, and the standards of colleges vary very much more than those of our universities. Some college standards are very high. Broadly, I think the generalisation is fair.

18 The number of engineers graduating per year in the United States is declining fairly sharply. I have not heard an adequate explanation for this.

19 The latest figures of graduates trained per year (scientists and engineers combined) are roughly U.K. 13,000, U.S.A. 65,000, U.S.S.R. 130,000.

20 One-third of Russian graduate engineers are women. It is one of our major follies that, whatever we say, we don't in reality regard women as suitable for scientific careers. We thus neatly divide our pool of potential talent by two.

21 It might repay investigation to examine precisely what education a hundred alpha plus creative persons in science this century have received. I have a feeling that a surprising proportion have not gone over the strictest orthodox hurdles, such as Part II Physics at Cambridge and the like.

22 The English temptation is to educate such men in sub-university institutions, which carry an inferior class-label. Nothing could be more ill-judged. One often meets American engineers who, in a narrow professional sense, are less rigorously trained than English products from technical colleges; but the Americans have the confidence, both social and individual, that is helped through having mixed with their equals at universities.

23 I have confined myself to the University population. The kind and number of technicians is another and a very interesting problem.

24 The concentration of our population makes us, of course, more vulnerable also in military terms.

25 There is one curious result in all major industrialised societies. The amount of talent one requires for the primary tasks is greater than any country can comfortably produce, and this will become increasingly obvious. The consequence is that there are no people left, clever, competent and resigned to a humble job, to keep the wheels of social amenities going smoothly round. Postal services, railway services, are likely slowly to deteriorate just because the people who once ran them are now being educated for different things. This is already clear in the United States, and is becoming clear in England.

CPSIA information can be obtained
at www.ICGtesting.com
Printed in the USA
BVHW031629140319
542654BV00003B/214/P

9 781614 275473